Sponsored Technology Workshops

Held at the 2023 AIChE Spring Meeting and 19th Global Congress on Process Safety

Houston, Texas, USA
12-16 March 2023

ISBN: 978-1-7138-7939-8

Printed from e-media with permission by:

Curran Associates, Inc.
57 Morehouse Lane
Red Hook, NY 12571

Some format issues inherent in the e-media version may also appear in this print version.

Printed with permission by Curran Associates, Inc. (2023)

For permission requests, please contact AIChE
at the address below.

AIChE
120 Wall Street, FL 23
New York, NY 10005-4020

Phone: (800) 242-4363
Fax: (203) 775-5177

www.aiche.org

Additional copies of this publication are available from:

Curran Associates, Inc.
57 Morehouse Lane
Red Hook, NY 12571 USA
Phone: 845-758-0400
Fax: 845-758-2633
Email: curran@proceedings.com
Web: www.proceedings.com

TABLE OF CONTENTS

How Knowledge Automation closes skills and knowledge gaps in yourworkforce

Clint Jeffus

The Skills Gap is a challenge faced by every manufacturer today. And it is growing. Plants are challenged in recruiting people with the right experience, onboarding them and capturing/transferring knowledge from experienced people before they leave. The traditional way of onboarding new hires and relying on a few SMEs (Subject Matter Experts) on critical operations is not working.
Join to see a deep dive into a Knowledge Automation platform for field operators and understand how this can close the knowledge and skills gaps in your plants. Knowledge Automation is proven to solve this challenge, leading to safer, more productive operations. We will cover real case studies and present data on how leading companies around the world have successfully implemented the platform.
If you are in operations, maintenance, workforce development, training, process safety or other relevant plant-based roles, this workshop will provide you with practical solutions that are easy to implement in your plants.

Skilling up towards Sustainability

Chris Cogswell

In this workshop we will discuss gaining skills in sustainable engineering for current engineering employees and new graduates, to ensure they are ready for the challenges companies will face in the years ahead. We will focus on Knovel, a tool all AIChE members have access too, as well as discuss important concepts in sustainability engineers will need to know to become leaders in this space.

Process Simulation using Symmetry

Victor Quiroga

During this workshop we would like to present how using one process simulation integrated platform (Symmetry) will enable you to: Reduce applications cost & engineering time by adopting one platform to fully simulate and analyze pressure relieving systems; prevent any off-spec, human errors, design errors and avoid any penalties; and allow you to achieve a zero emission goal.

Pressure Relieving Systems simulation workflows is often distributed among multiple applications, each application takes care of a piece of the calculations and then at some point will have to share the results with other applications that need them. This distribution makes it challenging to keep data and calculations of all components in synch.

To further complicate things, ?the software market offers a specialized applications or software packages for each process engineering task? and yet each software package relies on data calculated by others and this makes it challenging to keep the data consistent?. Each time there is a data transfer there is potential for (1) mistakes in manual data transfer, (2) data loss in translation, and (3) ? Engineering teams to be working on different versions of the applications?.

Traditional process engineering simulation workflows challenges: difficult to keep the large amounts of data consistent and in synch?, time-consuming, and dangerously error-prone, and the process is unmaintainable and difficult to keep up to date.

Symmetry will be one platform to:

Model your full distribution and gathering system, Steady State & Dynamic Analysis, PIONA characterization, Overpressure analysis, Radiation and dispersion
studies, and more capabilities that will eliminate the need to use different applications and introduce more comfort and ease of use to users. The workshop will also show flexibility in using this platform on desktop and cloud.

And finally, we will discuss the outcome of one case study and evaluation performed using Symmetry.

How Are You Managing Risk Today and Into the Future?

Mark Breese

Risk Assessments are fundamental to managing and mitigating risks and hazards and therefore enabling a safe and effective workplace environment for your workforce and contractors. Traditional risk management techniques are being augmented with technologies which utilize historic process knowledge to create a more predictive and dynamic mitigation strategy. Additionally, organizations are exploring the use of robotics to reduce the likelihood of exposure for operations personnel. All companies considering integrating robotics technology into their facilities should be accustomed with the risk assessment process.

P&ID's and Information Management - How they can play together

Stephanie Nelson

An unfortunate trend in today's workflow is that once a P&ID has been released for construction, the integrity and data starts to get lost. As a manager you could pay upward of $2,000 for an updated P&ID sheet and still end up with a piece of paper that does not achieve your end goals. In this workshop we can help you understand how to create a value-added P&ID that achieves ease of use and provides a holding place for PSI which in turn will create ownership for the PSI and PSM data at your facility and improve your auditing success. Most facilities information retrieval is sporadic and forces you down a winding rabbit hole of document accessibility. The path usually followed is very time-consuming and confusing. We can show you a P&ID with enhanced data storage capability to solve your information gaps and create successful information management retrieval. When information access and maintainability becomes simple, it causes your personnel to take ownership of the information management and its requirements. This workshop would benefit anyone that needs to access and maintain PSM PSI for their facility and wants to have a simple solution for information management.

PHAs, Bowties and Barrier Management

Paul Tours

Many thousands of process hazard analyses occur each year. They focus on the identification of hazards, their causes, consequences and ask the question, 'Are we safe enough?' Understanding PHA results can be challenging for managers and rank and file alike. Leveraging Bowties to explain PHA results is becoming more and more common. Ensuring barriers or safeguards identified are working as expected is the focus barrier managements systems. This workshop explores the advantages of integrating PHAs, Bowties and Barrier Managements systems.

Leveraging Technology Advancement to enhance process safetyperformance in the energy transformation journey

Nancy Faulk

Leveraging technology advancement for continuous enhancement of process safety management is critical to meet the energy transformation requirements. The workshop is a great opportunity to learn about some key technology advancement and share interesting case studies around:

- Energy transformation and the implications on process safety
- Process optimization by digitalization of asset integrity management
- Complex pressure relief and flaring systems data management
- Process safety and process decarbonization

AI Process Optimization for Refineries and Petrochemical Plants

Abi Mukund

2 sessions (Note: the same content will be presented at each session)

8:30-9:15 AM: Session 1 - 30 min presentation and 15 min Q&A

9:15-10:00 AM: Session 2 - 30 min presentation and 15 min Q&A

AI is as big as a buzzword as technology today. How do you build the business case for AI at your plant? Come join us as we first define AI back to its scientific definition. Then, we will focus our attention on a very powerful form of AI called Reinforcement Learning and how it has been used in industry to optimize refineries and chemical plants to capture millions of dollars in bottom line value through direct control. If you are a refining or petrochemical expert and feel there is unrealized optimization potential at your plant, this is the right session for you.

AUTHOR INDEX

NOTES

NOTES